Hodder Cambridge
Primary Science
Foundation Stage

The Toy Box

Rosemary Feasey

This book belongs to

· ·

HODDER
EDUCATION
AN HACHETTE UK COMPANY

The children find an old toy box in the cupboard.

Look at all these old toys!

Talk about the toys in the picture.

Which toys can you name?
Are any of the toys like yours?

The box belonged to Dad.

Compare Dad's old toys to your toys.

Are any toys the same as your toys? Which are different? When did Dad play with these toys? Would he play with the toys now?

Rika and Yafi play a fun game using the old toy box ...

Faster! Faster!

Talk about pushing and pulling.

Who is pushing the toy box? What have the children made with the box? Does the box look easy or difficult to move? Why?

Talk about how pushing and pulling can make things move.

What could the children do to make the box move more easily?
Which other toys in the picture move by pushing and pulling?

The children think about tidying up …

I have some roller skates like these in my bedroom.

Talk about how the toys in the picture move.

Which toys use wheels? Can you point to toys that do not use wheels?
Say which toys you push or pull to make them move.

Talk about how magnets work.

Where is the magnet on the fishing rod? How is the fish attracted to the magnet? Does the magnet fishing rod pull or push the fish?

Talk about what makes the toys work.

What makes the torch work? What toys can you spot that use batteries? Which toys make a sound with or without batteries?

Some of the old toys use batteries.

Talk about things that need batteries to make them work.

Why do you think the Walkie Talkies do not work? How does Yafi know they are not working? What could he do to make them work?

Adi finds some *really* old toys.

Those shadow puppets belonged to your Grandad!

Talk about how shadows are made.
How are Adi and Rika making the shadow on the wall?
What would happen to the shadow if Rika took away the torch?

The children do a puppet show for Dad.

Talk about the toys the children are using for the show.

Which toys are the children using? How are they the same or different to toys today? What is Yafi using to make the sounds?

The children decide to play all the instruments.

Let's make really loud sounds!

Compare how each of the instruments is played.
Which instruments are blown, hit or plucked? What kind of sounds will the instruments make?

Talk about loud and soft sounds.

How can the children make the sounds louder? How can they make the sounds softer? Can you remember all the toys the children played with?

Do you use a push or a pull to move these toys? Circle the correct word.

push	pull
push	pull
push	pull

 Join the instrument to the way it is played.

| blow | pluck | hit |

✓ the toys that use a battery to work.

Draw a line from the toy to its shadow.

Circle the correct end to the sentence.

To make a loud sound
on this drum, I would
hit it gently hit it hard.

To make a soft sound
on the trumpet, I would
blow it hard blow it gently.

To make a soft sound
with the guitar, I would
pluck it hard pluck it gently.

 Draw a toy for each sentence.

This is a toy that
makes a sound.

This is a toy that
has a magnet.